Axel
Tandem Truck

Coretta
Combine

Gertie
Dusty's 4-Wheeler

Pete
Gramp's Pickup

Riggins
Application Truck

Turbo
Dad's Pickup

Little Tricia
Mom's Utility Tractor

Big Travis
Tractor

To cows all over the world.

—DS

OUR FAMILY FARM
COWS ON THE MOOOVE!

WORDS & PICTURES BY

Dana Sullivan

North Dakota Farmers Union

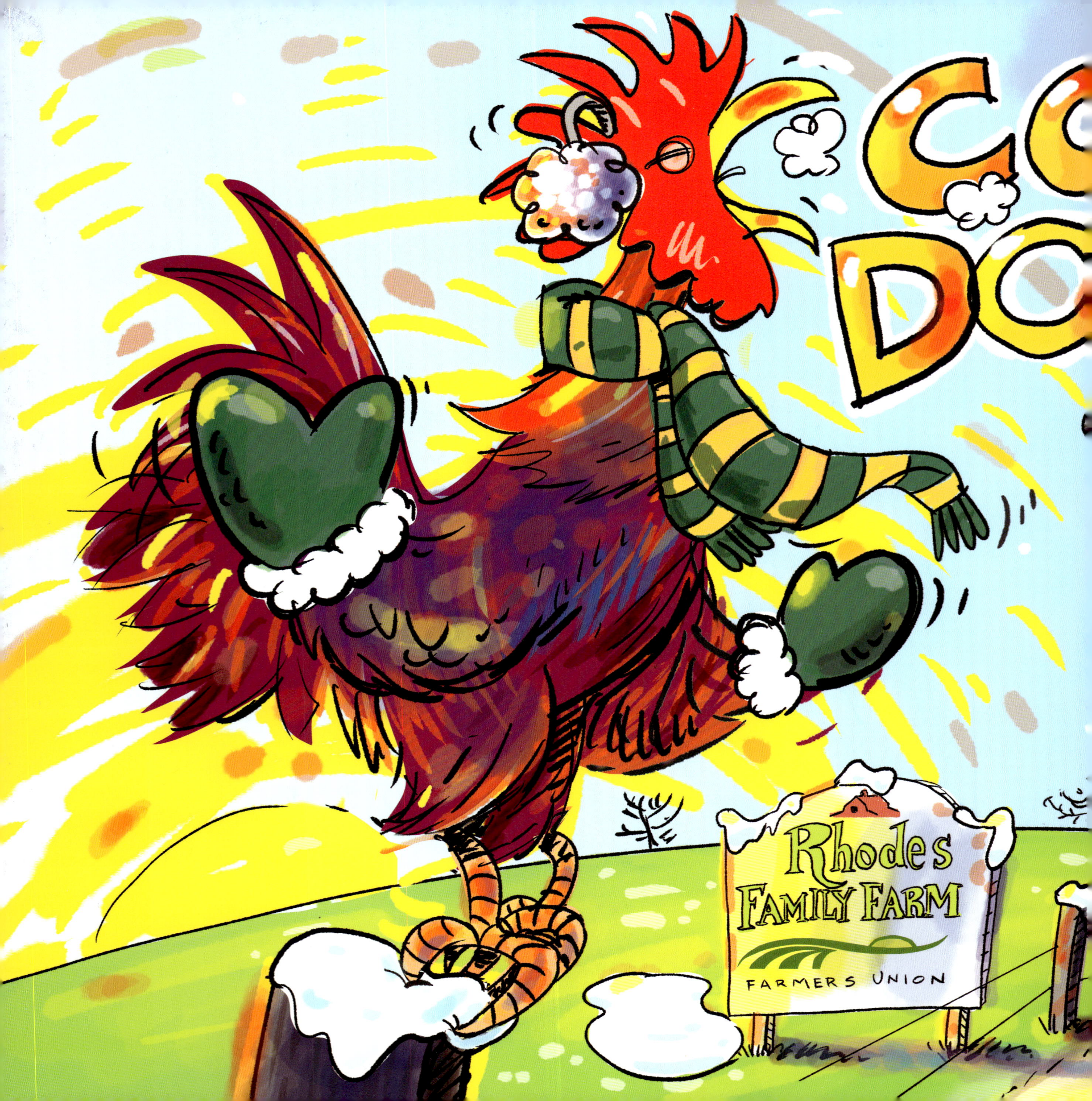
Rhodes
FAMILY FARM
FARMERS UNION

CK-A-ODLE-DOO!!!!

It's spring, and calving season on the Rhodes Family Farm. Red Rooster lets everyone know there's work to be done!

There's a lot to do before the cows have their calves. After a big breakfast, Dusty checks on her prize-winning cow, Daisy.

CLANKA CLANK!

Then she and Gertie help out in the tractor shop.
Gramps fits Big Travis with his grapple fork for moving bales of hay.

Axel is trying on his feed chute,

while Dad gets Riggins and Little Tricia ready for **calving season!**

Mom and Grams are installing cameras in the calving barn to keep an extra-special watch on the first-time mama cows.

When all the machines are in tip-top shape,
Dusty, Gertie and Rocky head out to the cow pasture.

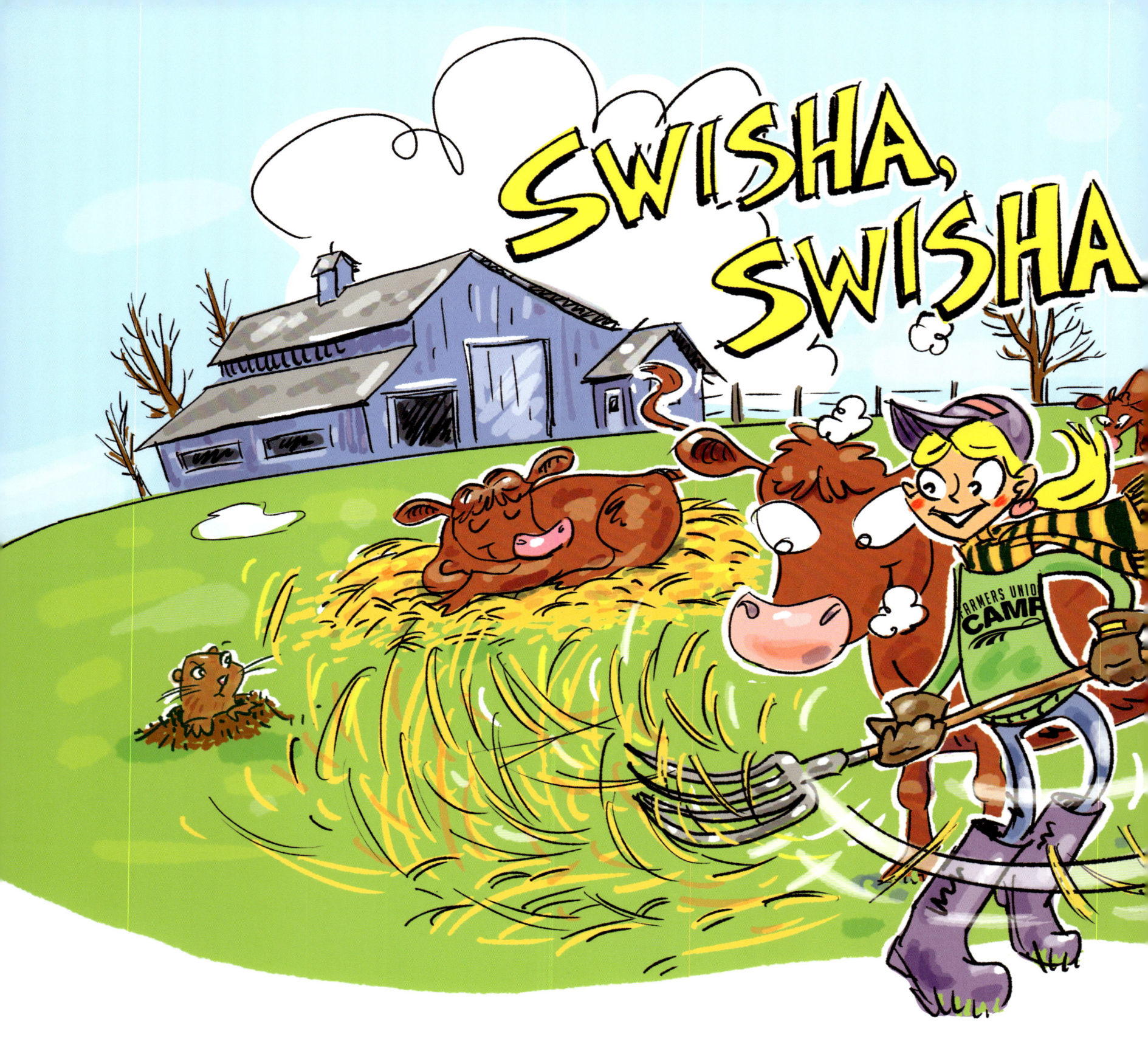

The cows are getting ready to have their calves.

AHHH!

They need fresh, clean straw for their beds.

There's so much work to be done! The cows are hungry and thirsty.

And so are the machines!

At night, Gramps checks on the cows in the barn with his phone.
Dusty and Rocky have one last chore before bed.

They need to check on Daisy one more time!

Late at night, Daisy wanders off.

Rocky senses something is up and investigates.

Daisy had her calf! Is that Rocky coming to congratulate her?

Oh no! It's a coyote!

The coyote is looking for a midnight snack.

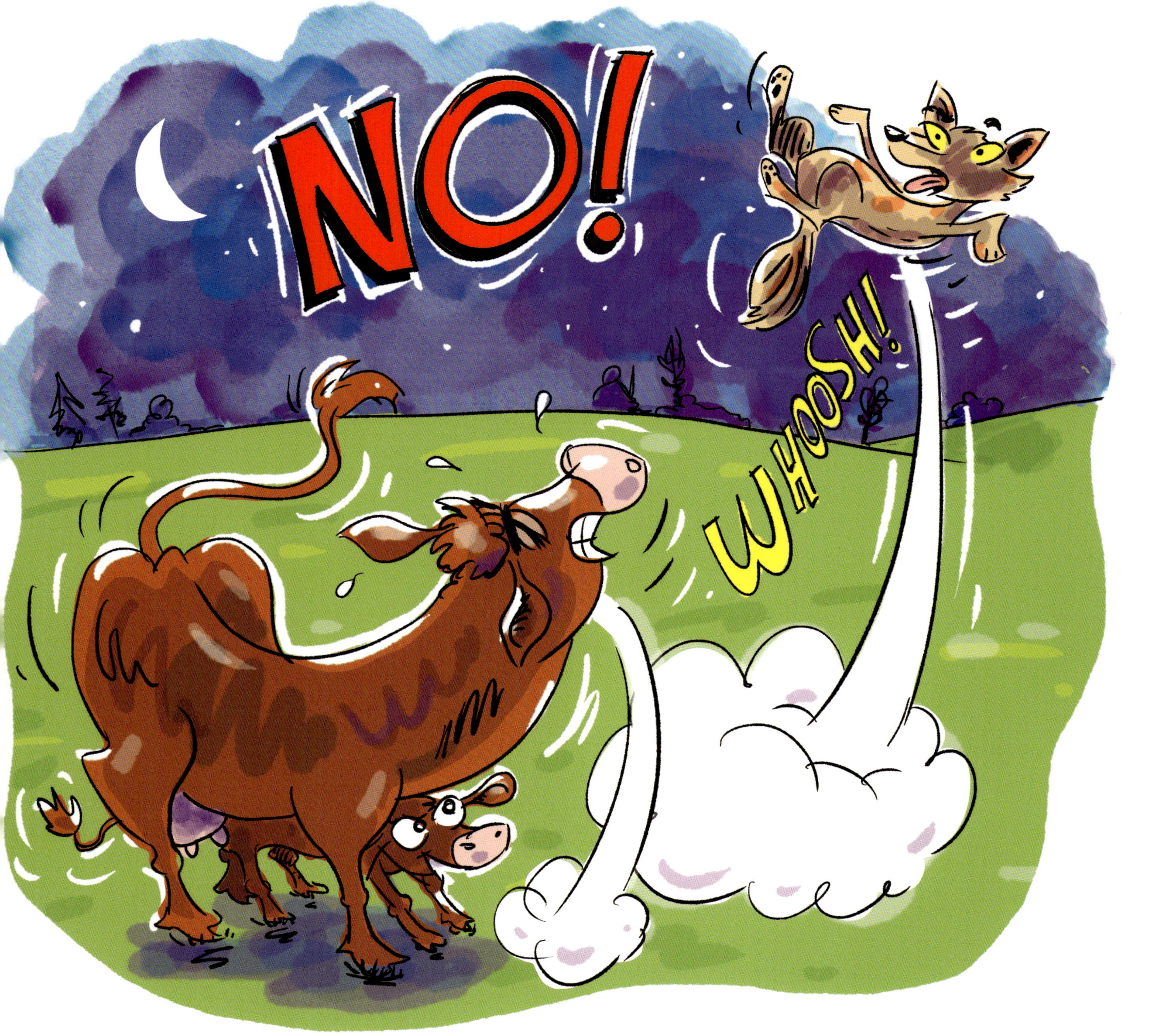
NO!
WHOOSH!

Another coyote?

No, this time it really is Rocky!

Rocky leads Daisy and her calf back to the calving pasture.

HOORAY!

"Good boy, Rocky! Good girl, Daisy! What a beautiful calf!"

It's the day of the big mooove! Dad fuels up Turbo for the trip.

Big Travis helps the cows load up to ride to the summer pasture.

They made it! All the farm families help the cows and calves settle into their fresh, green home.

After all their hard work, it's time to celebrate with **neighbors** and **friends**!

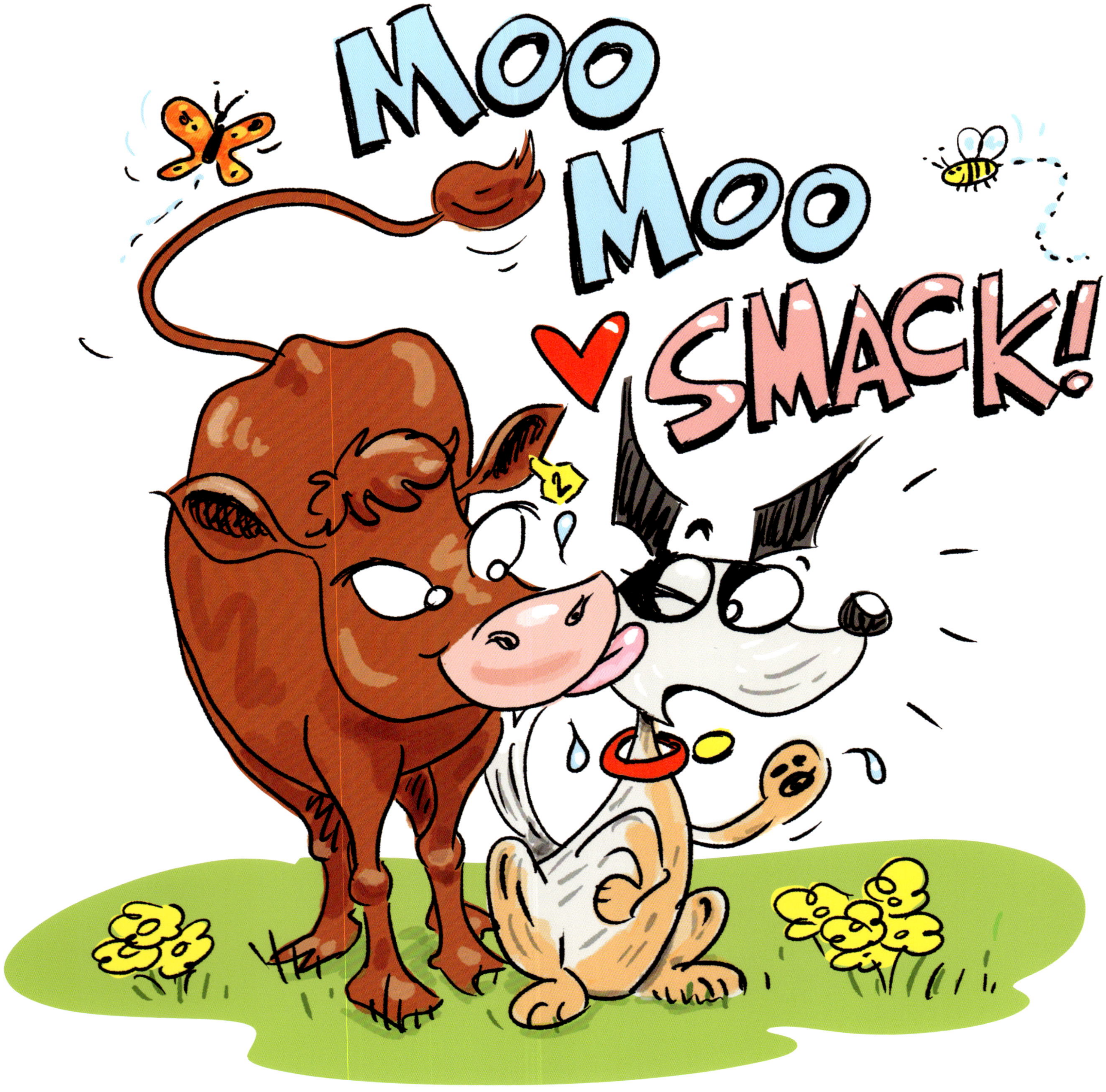
MOO
MOO
SMACK!

Copyright © 2021 by North Dakota Farmers Union Foundation.

All rights reserved. No part of this book may be used or reproduced in any manner whatsoever without express written consent of the publisher, except in critical reviews and articles. All inquiries should be addressed to:

North Dakota Farmers Union Foundation
1415 12th Ave. SE
Jamestown, ND 58401
www.ndfu.org

Library of Congress Control Number: 2021909413

ISBN: 978-0-578-91271-4

Printed in the United States of America.

This instructional text/artwork was commissioned by the North Dakota Farmers Union Foundation as a work for hire by its creator Dana Sullivan.

10 9 8 7 6 5 4 3 2 1

MORE inFARMation!

A cow has 32 teeth and will chew up to eight hours a day.

Cattle have four compartments in their stomachs. They digest plant material by repeatedly regurgitating it and chewing it again as cud.

Cows can sleep standing up.

A young female cow who has never had a calf is called a heifer. Heifers usually give birth in a calving barn, while older cows give birth in a pasture.

A newborn calf is commonly tagged in the ear. Each tag has an individual number which helps ranchers pair the mother to her calf.

Many ranchers have cow-calf operations. They keep a herd of cows to produce calves.

The meat from cattle is called beef. The average American eats about 65 pounds of beef a year.

COWA-BUNGA!

North Dakota Farmers Union is a grassroots organization committed to the advancement of family farm and ranch agriculture and quality of life for people everywhere through member advocacy, educational programs, cooperative initiatives, and insurance services.

If you like this book, go to **ndfu.org** to purchase more "Our Family Farm" stories.

Dana Sullivan was born in the big city, but now lives in tiny Port Townsend, Washington, with his sweet wife, Vicki, and their barky dog, Bennie, surrounded by farms and forests. Dana's favorite color is dog and his favorite vegetable is peanut butter. Dana would love to see photos of your prize cows, cats, dogs or goldfish. See his contact info and other books at **danajsullivan.com**.

Dad

Mom

Dusty

Grams

Rocky

Gramps